Marcelo Olimpio Gomes

# Facebook as a Tool for Popularising Astronomy

Marcelo Olimpio Gomes

# Facebook as a Tool for Popularising Astronomy

## A study on the potential of Facebook as a tool for popularising astronomy

ScienciaScripts

**Imprint**
Any brand names and product names mentioned in this book are subject to trademark, brand or patent protection and are trademarks or registered trademarks of their respective holders. The use of brand names, product names, common names, trade names, product descriptions etc. even without a particular marking in this work is in no way to be construed to mean that such names may be regarded as unrestricted in respect of trademark and brand protection legislation and could thus be used by anyone.

Cover image: www.ingimage.com

This book is a translation from the original published under ISBN 978-613-9-74555-5.

Publisher:
Sciencia Scripts
is a trademark of
Dodo Books Indian Ocean Ltd. and OmniScriptum S.R.L publishing group

120 High Road, East Finchley, London, N2 9ED, United Kingdom
Str. Armeneasca 28/1, office 1, Chisinau MD-2012, Republic of Moldova, Europe
Printed at: see last page
**ISBN: 978-620-6-29689-8**

Copyright © Marcelo Olimpio Gomes
Copyright © 2023 Dodo Books Indian Ocean Ltd. and OmniScriptum S.R.L publishing group

*To my beloved parents, Ataíde Olimpio and Joana Darc Gomes, for their unconditional love and endless incentives. To my beloved siblings, Diego Divino Olimpio Gomes and Valdênia Olimpio Gomes das Dores. To my beloved nephews, Diego, David, Ana Karolina, Valdênia Eduarda and Nicole. To all my dear relatives and friends.... Especially to Pr. Reginaldo Pinto Almeida, for accompanying me in difficult times, and who often did more than I expected, even though he was not a biological father, acting as such. To Prof. Dr Ademir Cavalheiro for his support, encouragement and advice at times when I almost gave up the course, as well as discussions on subjects that were often not part of the subjects covered in the classroom.*

# Acknowledgements

To God, for the gift of life and for all the achievements in my life.

To Prof Dr Ademir Cavalheiro, for his companionship, encouragement, dedication and competence to teach Physics. My eternal gratitude to Prof Dr José Luís Petricelli Castineira. To Prof. Dr Marcos Daniel Longhini for the continuity and conclusion of this monograph, to Profª . Drª . Diva Souza Silva for participating in the examining board. I would also like to thank the other participants of the board and the professors I had during my undergraduate studies at UFU, for the opportunity to learn and socialise.

To all friends of Morrinhos - GO, for the corroboration in overcoming difficulties that I faced since my childhood and for the victories achieved until the present moment.

To the teachers of Colégio Estadual de Goiatuba (Goiatuba - GO). To the teachers of Colégio Estadual Sylvio de Mello, Colégio Estadual Xavier de Almeida, Escola Estadual Drª . Gertrudz Lutz, in Morrinhos - GO, where I did not have teachers, but friends, whom I will carry for eternity, especially Professor Maria de Fátima, Professor Sandi Rezende Pereira, Professor Genice Luiza de Souza, among others who always encouraged me to continue my studies.

To my undergraduate friends: Marcos, Luismar (Ching-ling), Nilmar Camilo (Beraba), Bruno Messias, Pablo Marques (Baino), João Lucas (Tico), Stanley Lima, among others, for their friendship and companionship throughout the course.

To Maykell Figueira (Maykelsong), who introduced me to the world of bonsai, in which I learned a lot from his tips, as well as discussions that enriched me a lot. This made possible a personal realisation, which came at a time of fundamental relevance, since at the beginning it was simply a *hobby*, but because I liked so much to mess with plants (a fact that in adolescence was one of the things I liked to do the most), they were taking on a proportion that I did not expect. When I realised, I had more than 3.000 plants, and with that I saw the possibility of becoming an entrepreneur in the area, thus founding, on March 28,

2015, Viveiros Samai (VSI), which has as its central objective the Revitalisation of Springs (RN), Degraded Areas (AD), Permanent Preservation Areas (APP) and Legal Reserve (RL), in addition to the formation of orchards and, of course, bonsai of the most varied species of plants, both fruitful or not and concomitantly, work with native seeds of the Cerrado.

To the partner, friend, *brother*, companion Nilmar Camilo, who in rare exceptions, cannot help me, however, "The man of many friends must show himself friendly, but there is a friend closer than a brother." (Proverbs 18:24), that is, having a friend, a true friend is a gift from God. So we must not underestimate our true friendships. We must cultivate them, we must work on them so that they always grow and so that they are not lost during our journey, because old friendships must be very careful, even because friends are not found anywhere.

To the friends from dance, José Elias, who helped considerably in the improvement and understanding of figures and, consequently, to dance better; to Altamirando Neto Colombo for believing in the creation and ideas for the *Pratic Dance* group and to the other components: Daniela, José Adriano, Camila Zanetoni and Kris, among many other friends I made in dance.

To Leandro Teodoro, who I was able to have as a dance teacher, and who became an invaluable friend, with whom I found something more than just dancing, but an idea of what it really is and how it can go beyond being a physical activity, and help as a therapeutic activity.

I could not leave out the Barbed Wire Republic, where we had excellent moments of relaxation and parties. Its motto was "República Arame Farpado - Cerca o gado e fura o couro", a slogan created by Mr Beraba and Mr Baiano, with their stories, jokes and unforgettable moments. We had an illustrious non-resident, but who became part of the republic, our dear friend Tico, philosophical discussions coming from our Messiah, Bert's remarks and Alex's comments, Maykelsong's Socratic maieutics and sharing the room with Mr Beraba.

Rafael, where we were able to watch good films and some not so good, "Old Boy", which frustrated us as a film, but we discussed it until the wee hours, among others.

No matter how difficult the walk is, it was long days to get here, dark moments arose amid indecision, adversities will always arise at all times. Overcoming, however, requires objectivity and even sacrifices, because procrastination can be palliative and opportunistic, but it does not solve anything, because it postpones the inevitable, with postponements, with delays, with postponements, transfers responsibilities, in short (...) attitudes, if done when necessary and in the right time....) attitudes that, if done when necessary and at the right time, that is, if done at that moment, would not result in stress, a sense of guilt, loss of productivity and shame in relation to others, for not fulfilling our responsibilities and commitments, which unfortunately garnishes factors such as anxiety, low self-esteem, anguish, dissatisfaction, restlessness, unnecessary consternation and a self-destructive mentality, lacking coherence in knowing how to live, which unfortunately can reach the picture of depression, this being the disease that is seen as a true mental disorder, highly debilitating, considered by many as the "evil of the century".

More or less days the victory comes, after long internal and external battles, weariness is overcome, disagreements disappear, strength is renewed. Finally, my friends, brothers, teachers, colleagues, etc., fill your minds with all that is good and deserves praise, that is, all that is true, praiseworthy, honourable, lawful, worthy, right, pure, pleasant and decent.

To all my respect and sincere thanks.

*"The mind that opens itself to a new idea will never return to its original size." (Albert Einstein) (Albert Einstein) "The important thing is not to stop questioning. Curiosity has its own reason for existing. One cannot but feel reverent when contemplating the mysteries of eternity, of life, of the marvellous structure of reality. It is enough that one just tries to understand a little more of this mystery every day. Never lose a holy curiosity".*
*(Albert Einstein)*

*"The wise man never says everything he thinks, but he always thinks everything he says."*
*(Aristotle)*

*"Remember that sleep is sacred and feeds the awake time of living with horizons". (Beto Guedes)*

*"Friendship doubles the joys and divides the sorrows".*
*(Francis Bacon)*

*"The art of living is more like fighting than dancing, in that it is ready to face both the unexpected and the unforeseen and is not prepared to fall."*
*(Marco Aurélio)*

*"Science is the progressive approximation of man to the real world". (Max Planck)*

## Summary

The teaching process is constantly changing and always seeks new solutions to facilitate the understanding and abstraction of concepts. With the emergence of new Information and Communication Technologies (ICT), social relations have changed in order to change the way the individual relates to the environment. In this scenario, Facebook is a social network that enables interaction between various social and cultural segments present in society. In Brazil, there are more than 136 million users who use Facebook. Despite its potential, this is a technology that is still little used for educational purposes. This social network presents some resources that can be used, for example, for the Popularisation of Science, especially Astronomy, our focus of investigation. For this work, our central objective was to analyse the potential of social networks, especially Facebook, with regard to the popularisation of Astronomy. As a methodology, we investigated which are these groups and to which institutions they are linked, the number of people they reach, the type of publication they offer and to which sub-area of Astronomy they are most dedicated. The data were organised into the following categories: Universities, Centres, Colleges, Museums, Associations and Mixed. We obtained a total of 427,014 users and 1246 publications collected for the period from 12 May to 16 October 2017, grouped into the following subcategories: Dissemination, Interaction, Sharing and Sales. The data showed that most of the groups are linked to some institution, which favours the origin of the publication made. The most common type of publication is the sharing of photos, news, disclosures, visits, videos and events, which together account for 67% of the publications made in the groups. The area that is most dedicated is Cosmology, but we noticed a trend in Astrophotography. Based on the results, we conclude that despite being restricted to Facebook users, this tool can and is a complementary educational tool, which makes it possible for people who do not attend school environments to obtain this scientific information.

**Keywords:** Social Networks; Popularisation of Astronomy; Facebook.

## Summary

# CHAPTER 1

# INTRODUCTION

Science, together with technology, has provided notable advances to society, whether developed or developing in areas such as health, education, socio-economic and socio-cultural, which, in a way, have eliminated the geographical isolation of countries, that is, shortened distances, promoting the so-called globalisation. This has been characterised by a set of transformations in the economic, social, cultural and political spheres, being an increasingly decentralised phenomenon, which is not under the control of any group of nations and even less under the control of large companies (GIDDENS, 2013).

An example of this advance can be the fact that society is on the *Web*, a word of English origin, meaning "web" or "network". The meaning of *web* took on another meaning with the emergence of the *internet, coming* to designate the network that connects computers all over the world.

This worldwide computer network (**International-Networking** or just InterNet) emerged in the mid-1960s in government military institutions in the United States. Today, there is an extensive range of information resources and services, such as the interrelated hypertext documents of the *World Wide Web* (WWW), *peer-to-peer* networks and infrastructure to support electronic mail (e-mails). The Internet has a major differentiating factor from other technologies, particularly in terms of interactivity. Networking involves, in addition to collaboration, interactivity and a dialogical process.

The democratisation of access to technologies has contributed to their dissemination, as well as to the use of new strategies, not only in relation to teaching, but for interactive exchanges between individuals (users), promoting interpersonal relationships.

It was J. A. Barnes, in the mid-20th century, who first used the term Social Networks (SN) to systematise the patterns of ties in society at that time, later incorporated into contemporary concepts, and then came to be composed of a complex set of relationships between members of a social system of different dimensions, i.e. well-defined social groups and categories (tribes, families, gender, ethnic group) (CREATIVE, 2016).

SRs have been the subject of research, such as those that have investigated their potential as a pedagogical tool (FERREIRA, CORRÊA and TORRES, 2013; CRUZ, 2014; GOMES, 2013).

With Social Networks, there is a need for the teacher to seek new teaching strategies to make use of pedagogical resources available in cyberspace, such as those that use Web 2.0 tools. On this topic, Gonçalves (2010) emphasises that:

> "Web 2.0 tools, such as RS, provide many opportunities to create an effective, efficient and engaging learning environment. Innovation, collaboration, interaction, sharing, pro-activity, participation, critical and reflective thinking, are some of the key words of using Web 2.0 in an educational context". (GONÇALVES, 2010, p. 3).

In this SR structure, social actors are characterised more by their relationships than by their attributes (gender, age, social class). Tomaél (2005) reaffirms the idea that SRs are present in the lives of most individuals in society, that is, in different social classes, educational levels, among other aspects. It is no different, since the number of people connected virtually through various communities is increasing, as verified by the Statista *website* (2017), which made a survey containing a list with the RS with the largest number of active users in the world, measured in millions of users, as shown in Figure 1.

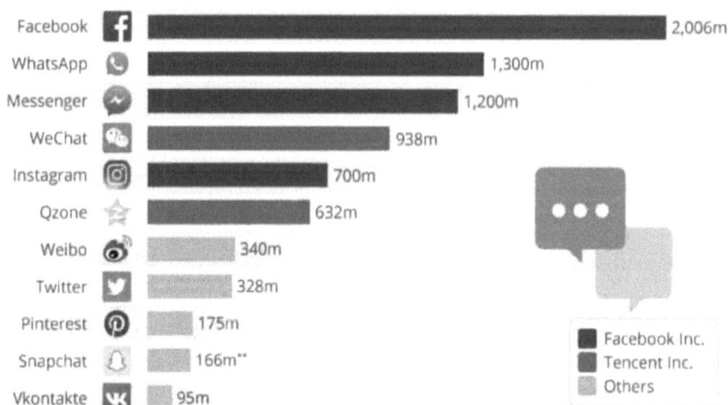


**Figure 1: The 11 social networks with the highest number of users.**
Source: Statista, 2017.

Brazil is one of the countries where the number of Facebook users is growing the most, according to an analytical report released by SocialBackers (2012). Recently, Brazil lost to India the position of "second largest", in number of people connected on Facebook (PERON, 2016).

Lévy (2007) and Castells (2004) call this phenomenon "network society" or global society, since the whole world is affected by the processes that take place in global networks, reducing distances and bringing people with common interests closer together. Each individual is an agent that disseminates information and, simultaneously, a node in the vast web of information, as visualised in the scheme, where the red dot represents greater interconnections between the SRs, since, if the individual has no connections, he will be more distant from the SRs as shown in Figure 2.

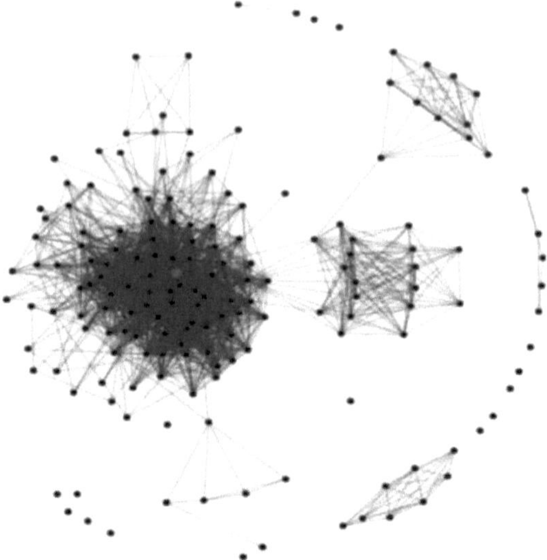

**Figure 2: Connections between the various networks.**
Source: Darwin Peacock, 2009.

It is in this context that the concept of "cyberspace" arises, which Monteiro (2009) defines as a communication space opened up by the interconnection of the *web* and computer memories, that is, it is the space in which digital information circulates, allowing people to build and share collective intelligence.

Nevertheless, it is common to hear the term "Social Media" (SM), which despite being analogous to SR, however, is a virtual space where people are connected in groups (network), such as communities (Twitter, Forums, Chat Rooms) that are used by Internet users as a place to obtain information of the most varied types. MS are platforms used to communicate news, announcements, messages, events etc., such as Facebook, LinkedIn, MySpace, Twitter, Blogger, Wordpress, Sonic and Youtube.

Such networks can operate at different levels, and can be represented by a diagram, containing fundamental characteristics of openness and porosity, enabling horizontal and non-hierarchical relationships between participants, and therefore allowing relationships regardless of the RS used. They can be separated into relationship, professional, community and political networks, among other types, as shown in the diagram in Figure 3.

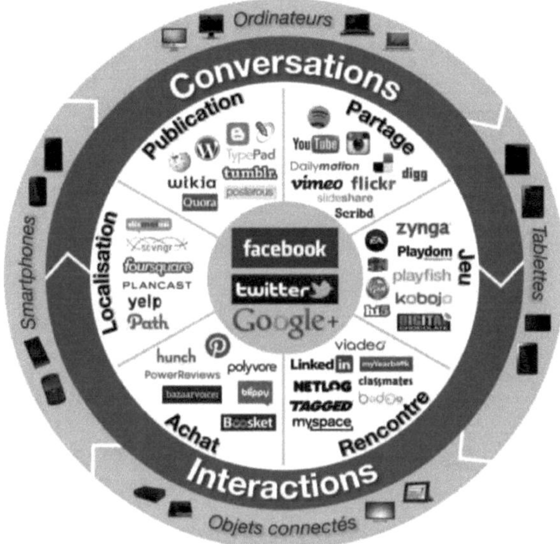


**Figure 3: Schematic representation of the most diverse Social Media.**
Source: Pinterest (2016).

Aquino and Brito (2012) point out that, among the various RS available, the most used today and which has been gaining more and more users around the world is Facebook, as it allows great interactivity among its users, which according to Calipo (2008), is a characteristic trait of contemporary youth.

As previously pointed out, this need to master the technological tools increasingly used in contemporary society is already a reality of education professionals, as Motta and Gava (2010) and Martins (2010) point out, as these professionals recognise the changes generated in the educational context and the requirements presented for new forms of learning.

As a result of these changes, technology, in general, is constantly changing, and transforms the way of teaching and learning (PAPERT, 1993), offering greater versatility, interactivity and flexibility of time and space in the educational process. In a certain way, the RS have become part of people's daily lives, and this is a reality. More than entertaining, they can become valuable interaction/intervention tools to assist in the dissemination of science, for example. In view of this, it is possible to ask: what possibilities would Facebook provide to the process of popularising science?

# CHAPTER 2

## FACEBOOK: Historical Development

Facebook was created in 2004 by American Mark Elliot Zuckerberg, Dustin Moskovitz, Chris Hughes and Brazilian Eduardo Saverin, while still students at Harvard University.

Initially, use of the *site* was restricted to students at that university, but it quickly expanded to other Boston colleges, from the Ivy League, Stanford University, to Columbia University and Yale University. According to Arie Hasit, a friend of Zuckerberg:

> "We had books called Face Books, which included the names and photos of everyone who lived in the student dormitory. Initially, he built a website and put up two pictures of two guys and two girls. Visitors to the site had to choose who was **hot** and according to the votes there would be a **ranking**, giving rise to the name of the app." CREATIVE (2017).

The original intention of Facebook's creators was to provide a virtual environment that would allow people who were not so popular among university students, the so-called *nerds, to* expand their circles of friendships and post, in addition to academic subjects, photos, videos and user profiles, with the aim of fostering relationships. Gradually, the site added support for students from several other universities, before allowing access for high school students and eventually for anyone aged 13 or older. Figure 4 shows the homepage of this social network, highlighting the possibility for any individual to create an account for use.

**Figure 4: Facebook homepage.**
Source: Facebook (2017).

The growing number of Facebook users made Zuckerberg gain notoriety, and in mid-2012, he made the following post:

> "As of this morning, there are more than 1 billion people actively using Facebook each month. If you're reading this: thank you for giving me and my small team the honour of serving you. Helping a billion people connect is wonderful, rewarding, and by far the thing I'm most proud of in my life. I'm committed to working every day to make Facebook better for you, and I hope that together one day we'll be able to connect the rest of the world." EBC (2012, online).

In November 2016, Facebook's Chief Executive Officer (CEO) announced that RS had reached 1.8 billion active users (getting closer to 2 billion, representing about 25% of the world's population of 7.5 billion), according to WordOmeters (2017). Of these, 1.2 billion use the app on *smartphones,* reinforcing the importance of RS. Data shows that the number of people using Facebook Live (via *smartphones* and *tablets*) to make live videos has increased fourfold since May 2016. In July 2017, Statista (2017) highlighted a *ranking of* Facebook users by country, with data revealed in Table 1.

**Table 1: Numbers of active users per country on the Facebook platform.**

| Countries | Active users in millions |
|---|---|
| India | 241 |
| USA | 240 |
| Brazil | 139 |
| Indonesia | 126 |
| Mexico | 85 |
| Philipinas | 69 |
| Vietnam | 64 |
| Thailand | 57 |
| Turkey | 56 |
| United Kingdom | 44 |

Source: Statista (2017).

Technologies, in general, evolve in such a way that it is possible that the current generation does not adapt with the technologies that come from this evolution. Facebook is a typical example of this, as mentioned earlier by Zuckerberg. In 2012, there were just over 1 billion people connected monthly. Currently, due to the improvement of technology, it has made people's lives easier and closer, that is, the phone has become more than just a simple equipment; with it we can listen to music, *download*, take pictures and share them on Facebook, giving it incredible milestones in the digital age, thus creating an expressive number of accesses, posts and shares of videos and images never seen before. The data on the types of actions carried out on Facebook are shown in Table 2.

**Table 2: Types of access and what users do when accessing Facebook.**

| | |
|---|---|
| Posting via Desktop. | 349.60 million |
| Posting via Smartphone. | 184.71 million |
| Post via T ablet. | 87.86 million |
| Videos viewed per day. | 8 million |
| Photos and Images posted. | 136,000 per min |
| Photo uploads per hour (Even the specialised photo site Flickr doesn't get that many photos: "only" 3.5 million per day, according to The Verge, and the total is 8 billion. | 14.58 billion |
| Shared publications. | 4.7 billion per day |
| Shared posts - "Like" button. | 4.5 billion times every 24 hours |

Source: Blogmidia and Facebook (2016).

Throughout this decade, the existence of Facebook has had a direct impact on news consumption habits, on the pattern of 'relationships' between people and organisations, and has increasingly influenced political and electoral decisions.

A few years ago, Facebook was fundamental for the expansion, dissemination and communication of some events, such as the "Arab Spring", "Indignados of Spain" and "*Occupy Wall Street*" (GOHN, 2016). In 2011, there were perhaps the first major social mobilisations, whose engagement occurred decisively through Zuckerberg's platform. In Brazil, we can mention the mobilisations of June 2013 (EBC, 2014). The network is so ubiquitous today that many users believe that Facebook is the internet itself.

For every 10 Brazilians connected, 8 are active on Facebook, and this gives us a single certainty: regardless of the segment to be addressed, a large part of the public is connected on Facebook. For this, we list the profile of the Brazilian public and the frequency with which they access Facebook, data present in Table 3.

**Table 3: Profile of Brazilians on Facebook.**

| | |
|---|---|
| Man. | 46% |
| Woman. | 54% |
| 13 to 17 years. | 9 million |
| 18 to 24 years. | 28 million |
| 35 to 44 years. | 29 million |
| 45 to 54 years. | 17 million |
| 55 to 64 years. | 10 million |
| Over 65. | 4.5 million |
| They use Facebook every day. | 2.1 million |
| Average time people spend on Facebook. | 22 minutes per day |
| Facebook hits peak on weekdays. | 1pm until 3pm |

Source: Blogmidia and Facebook (2016).

Earlier this year, on the eve of reaching the 2 billion user mark, Zuckerberg made two statements, one in March and the other on 22 June, saying that in the face of the new international context, the dilemmas of humanity, Facebook was called upon to change its mission. Initially designed to "connect people", it is now adjusted to "bring the world closer together" according to Facebook (2017).

# CHAPTER 3

# POPULARISING SCIENCE

Since antiquity, man has observed the universe around him, seeking greater understanding of the phenomena he perceives. The first records date back to the Prehistoric era, when drawings engraved on cave walls were used to communicate something to those who saw them. However, this way of transmitting knowledge only reached those who were present at the site (BRETONES, 1993). The exchange of knowledge between human groups in increasingly distant regions brought the challenge of creating means of communication that reached wherever someone was.

Over time, the fears and legends arising from representations gave way to systematic observation, which over the years were reinvented, introducing concepts that involved postulates, principles and laws in an attempt to explain something or some type of phenomenon not yet explained/studied. This search for the unknown gave rise to Science, a word originating from the Latin *SCIENTIA* meaning "knowledge", and *SCIRE* "to know, to know".

The means used to do science have evolved considerably, especially with the discovery of instruments capable of portraying information beyond that which could be captured by sensory organs. To this end, the senses have been expanded, telescopes have been built to see further and further and radio telescopes to "see" what the eyes cannot capture; microscopes have been created to examine the constitution of matter and particle accelerators to "observe" the interior of atoms, as well as offering the opportunity to know more and more about the cosmos.

A pioneering example of this is astronomy, which has fascinated human beings since the beginning. In 1609, with the construction of the astronomical telescope of Galileo Galilei (1564 - 1642), there was the beginning of the adventure of the sciences with the image in detailed drawings, which were tools of their arguments, that is, their theories, making these instruments increasingly sophisticated (NASCIMENTO, 2008).

In Brazil, Germano and Kulesza (2008, p. 7-25) state that the term "Popularisation of Science" gains new strength with the creation of the Department for the Dissemination and Popularisation of Science and Technology (DDPCT), a body linked to the Ministry of Science and Technology (MCT), whose main task is to formulate policies and implement programmes. In this same scenario, a decree by former President Lula created the National Science and Technology Week (SNCT), with the main objective of highlighting the importance of science and technology for people's lives and for improving the quality of education in Brazil. This annual event is funded by the Ministry of Science, Technology and Innovation (MCTIC) and takes place simultaneously in almost all Brazilian states.

The MCTIC proposes a different theme each year, leading the institutions participating in the event to develop educational and playful activities (lectures, films, videos, experiments, games, among others), showing the scientific and technological advances related to the pre-defined theme. The activities create a favourable environment for the exchange of ideas, promoting debates and stimulating the awakening of scientific vocations.

Much of the educational effort falls to the federal, state or local government. However, outside the academic environment, initiatives are emerging that allow the "ordinary" citizen to follow scientific progress, to be informed about what is happening, as well as to learn how to take better care of their health, or simply to kill curiosity and delight in their own discoveries.

These initiatives, ranging from science news to interactive science museums, through science magazines, formal and non-formal science spaces, have brought citizens closer to science, breaking paradigms, such as, for example, that science is generally seen as something distant, almost magical, restricted to ultra-secret laboratories, where divine beings work and immune to human passions (BRETONES, 1993).

For Reis (2002), more than telling the public about the charms and interesting and revolutionary aspects of science, science communication is the conveyance, in simple terms, of science as a process, of the principles established in it, of the methodologies it employs, revealing, above all, the intensity of the implicit social problems.

In order to train individuals for the dissemination/popularisation of science, Osmir Nunes, coordinator of the José Reis Centre for Science Journalism (NJR) at the University of São Paulo (USP), indicates an increase *in* demand for postgraduate courses in science journalism since 1992, not only because the current moment brings this demand, but also because science communication is beginning to constitute one of the so-called "careers of thc futurc" (BUENO and DIAS, 2008).

Another institution that understood the relevance of the need for scientific dissemination today was the Open Laboratory of Interactivity for the Dissemination of Scientific and Technological Knowledge (LAbI), of the Federal University of São Carlos (UFSCAR), which has been offering a specialisation course in scientific dissemination since 2006, as available on the website http://www.labi.ufscar.br/2017/07/01/caminhos/.

According to Germano and Kulesza (2008), science communication is believed to be more concerned with building a myth around science than with explaining to the public important aspects of the reality that surrounds it.

According to Germano (2011):

"The term "popularisation of science", considering science as a

> modern science, emerged in France in the 19th century as an alternative to the concept of vulgarisation, but did not achieve success in the French scientific community. Subsequently, due to the interest that it would have a greater impact, the current of communicologists (popularisers) prevailed, for whom the central idea was the transmission of messages and the processes that intervene in it. In the meantime, the popularisation of science achieved greater penetration among the British who, according to Mora (2003, p. 10), were more concerned with the product and the practical aspects than with the form". GERMANO (2011, p. 302).

From the issues raised and exposed above, we can understand that the concept or, rather, the definition we give to "popularisation of science" is not static, but dynamic, as it is renewed according to the pre-established relations between science and society and also according to the assimilation that one has about science and society.

Under this perspective of science and society, we will adopt the term Popularisation of Science to designate the means by which the dissemination of science knowledge occurs, although the concept of "dissemination" is understood as the act or action of disseminating, which, from the Latin, *Divulgare,* means "to make known", "to propagate", "to spread", "to publish", "to transmit to the vulgar", or even, "to make known", "to make popular". This sometimes causes conceptual confusion between the following terms: dissemination, literacy and popularisation of science, which according to FREIRE (1992):

> "It is in its structural whole that the word, in relation to the others, defines its meaning. Thus, based on the assumption that within the linguistic structural unit associative relationships are established that unfold between the significant fields of the various terms, we will proceed to an analysis of the words: vulgarisation, literacy, dissemination and popularisation. All of them are related to the issue of access to scientific knowledge, a common thread that links all the terms to the word science. We do not intend to discuss the concept of science, considering that, although controversial, it is quite consolidated". FREIRE (1992 p. 21)

For Martinez (1997), the popularisation of Science and Technology (S&T) is based on four pillars:

i. formal education;
ii. the mass media;
iii. multimedia programmes;
iv. interactive science centres.

Still according to Martinez (1997), formal education is basically associated with school environments, characterised by the presence of curricula.

With regard to the media (television, radio, newspapers, internet), the author emphasises the need to set up interdisciplinary groups, so that science communication (predominantly the responsibility of scientists and science journalists) enables a new use of the media.

With regard to interactive science centres, the same author comments that these places are recent spaces for public science learning, which could promote the inclusion of different sectors of society. Lastly, multimedia programmes would involve printed, audiovisual and computer-based materials, as well as other types of spaces that would provide cultural activities related to science.

To introduce the notions of formal, non-formal and informal education, Smith (2001) takes as a starting point the global education crisis, which was the topic of discussion at an international conference held in Williamsburg, USA, in 1967. The author relates how, since the 1960s, educational policy makers have realised critical elements in the education system, among them the limited flexibility of curricula and the inability of the formal education system to keep pace with socio-economic changes. In addition, strong tensions identified in the 1970s led organisations such as the World Bank to begin to make distinctions between formal, non-formal and informal education, and organisations such as UNESCO to move towards notions such as lifelong education and learning society.

On the other hand, Smith (2001) considers that criteria such as purpose, temporality, content, "delivery" system and control could be used to differentiate the three notions (formal, non-formal and informal education), in a less "administrative" logic than that suggested by Marandino (2004).

It should be emphasised at this point that even when there is no consensus among different authors on the definition of formal, non-formal and informal education, international discussions on education systems are permeated by a broad notion of education that can take place both inside and outside school in a temporal and spatial sense.

Statements given by specialists in the area of science dissemination have revealed, in recent years, the importance of considering the formulation of a National Plan for the Popularisation or Dissemination of S&T. Regarding this initiative, Melo and Diegues (2002) comment:

> "An initiative of this type will only be successful if there is broad popular participation [...] what exists today is an enormous healthy curiosity of society for scientific themes, whose satisfaction through this broad and rich mosaic of dissemination initiatives will allow their participation in a project of this nature". (MELO and DIEGUES, 2002, online).

The need to articulate actions at national level is instigating, a fact that has been expressed by the Brazilian Association of Science Centres and Museums (ABCMC), through the proposal of a National Programme for the Popularisation of S&T, supported by the following objectives (ABCMC, 2006)[1] :

> "To contribute to the structuring of a national system of popularisation and education in science that understands it as a process that aims to promote active exploration, personal involvement, curiosity, the use of the senses and intellectual effort in the formulation of questions and the search for solutions; that aims to offer answers, but above all to generate enquiry and interest in science.
> To promote the education of citizens capable of perceiving science in all its dimensions: as a source of pleasure, of transformation of the quality of life and of relations between men, but also as a historical and social process that, alongside the benefits, can generate controversies and offer risks to their lives, to the life of the community and to the environment and that must, therefore, be subject to constant ethical and political evaluation". (ABCMC, 2006, online).

The association's national programme contains five specific lines of action:

i. promoting the allocation of more budgetary resources at state levels;
ii. support for the development of science museums and centres and integrative networks linking these spaces;
iii. fostering partnerships between formal and non-formal education spaces;
iv. supporting events and practices that strengthen citizenship in relation to current S&T issues

[1] Information available on the company's website www.abcmc.org.br

and debates;

v. support for the development of wide-ranging science dissemination programmes and activities (NAVAS, 2008, p. 45).

The above considerations lead us to think about the relevance that museums and science centres have acquired for the area of S&T popularisation and the role they can play, in their own objectives and the possibilities, demands and needs of the environment.

On the national and global scene, science museums and centres have organised their exhibitions with the aim of improving communication with the public through a more interactive presentation (SILVA, 2006). In a way, Science Centres and Museums (CMC) emerge as privileged places where society can participate in the process of scientific and technological development in Brazil, which from 1870 onwards has been modernised, arising from the great incentive to agricultural production. According to Lopes (1997), this modernisation had an effect on science museums by breaking with the naturalist tradition that initially gave rise to them (NAVAS, 2008).

In addition to museums and spaces for scientific dissemination, it is also noted that the electronic media has great potential for scientific popularisation, as it is expected that the viewer/individual/reader will be able to recognise scientific content, even if this is not the purpose of the communication vehicle, as stated by Albagli (1996).

According to Mueller (2002), advances in science sometimes bring threats to humanity and the environment, but the role of science popularisation/dissemination has been crucial in raising awareness about the risks inherent in certain scientific discoveries. This awareness is due to the efforts of Governmental Organisations (GOs), Non-Governmental Organisations (NGOs), journalists, and also scientists, who despite the increase in the volume of news about science and the greater degree of understanding and scientific awareness of society, still often come up with conflicting versions about risks, which lead to uncertainty and insecurity.

In order to remedy these conflicting gaps, MS/RS are fundamental in the popularisation of science, as we believe that by disseminating (concept discussed above) some information on Facebook, thousands, better said, millions of people are connected (online) on the platform, and even with the most diverse purposes, can have access to such data or content.

# CHAPTER 4

# GENERAL OBJECTIVE

Considering that social networks have been developing in recent decades, that they can become tools in the dissemination of information, among them, scientific content and, finally, considering that knowledge in the field of Astronomy is often the target of news in the most different media, this work aims to *identify the potential of social networks, especially Facebook, with regard to the popularisation of Astronomy.*

## 4.1 Specific Objectives

Identify the number of Facebook groups that work to popularise astronomy;

Identify in which areas of astronomy these groups have focussed their action;

Identify the number of people who participate in these pages;

Identify which community/institution these pages are linked to;

Identify what resources these pages use as a tool to popularise astronomy.

# CHAPTER 5

# RESEARCH METHODOLOGY

This research has as its data source the network called Facebook. In order to carry out the research on such a network, it was necessary to use the researcher's own profile, since it is necessary to be connected to it as an active user.

The data analysis methodology was based on Bardin's Content Analysis (2011), which is organised in three stages: 1) pre-analysis, 2) exploration of the material and 3) treatment of results, inference and interpretation.

The first stage consists of analysing the records obtained on the pages of this social network, as well as the posts of its members. We obtained 145 pages related to Astronomy on Facebook/Brazil. These data are related to the consultation carried out in the period from 12 May to 16 October 2017, since there is a large number of publications, which could cause delays in their analysis.

The second stage of data analysis consists of the construction of coding operations, considering the clipping of texts into record units, the definition of counting rules and the classification and aggregation of information into symbolic or thematic categories. Bardin (2011) defines coding as the transformation, through clipping, aggregation and enumeration, based on precise rules about textual information, representative of the characteristics of the content.

Finally, the processing of the results, inference and interpretation, consists of capturing the content published in the group (publications, videos, images, events, polls, documents) and drawing from them the conclusion regarding the central objective of this research, as well as in relation to the specific objectives.

In this phase, all the material collected is cut into registration units, which will make up the publication units of each group, as well as the comments. From these publications, the keywords are identified, and each paragraph is summarised in order to carry out a first categorisation. These first categories are grouped according to related themes and give rise to the initial categories. The initial categories are grouped thematically, originating the intermediate categories and the latter are also agglutinated, depending on the occurrence of the themes, resulting in the final categories (SILVA and FOSSÁ, 2015).

The inspection of the Facebook website in October 2017 showed that this technological device provides the following tabs/tools:

i) **All:** covers all content of the referred search, such as: groups, pages, friends' posts, apps, videos that are related to the search;

ii) **Posts:** is divided into 3 sections: 1) When your fans are connected: shows when people who like your Page access Facebook content; 2) Post types: shows the success of various post types based on average reach and engagement; and 3) Top posts from Pages the user follows: reports post engagement from Pages the user is following;

iii) **People:** you should send friend requests on Facebook to friends, family and other people you know and trust. To add a friend, you search for them and send a friend request. If the person accepts, you will automatically follow them and they will follow you back, which means you will both be able to see each other's posts in your News *Feed;*

iv) **Photos:** you can share a photo or create an album with a collection of photos from a special time or place, such as a birthday party or holiday. You choose who can see your photos and albums. If someone tags you in a photo you don't want to be tagged in, you can un-tag them;

v) **Videos:** Users can add videos via a file from their computer, by adding directly from their mobile phone via Facebook Mobile or by using a direct recording feature from a *webcam.* In addition, one can "*tag*" their friends in the videos. This feature came about due to competition with MySpace. However, Facebook Video does not allow you to share videos outside of Facebook or download or export uploaded videos.

vi) **Shop:** intended for sale and purchase, making it more attractive to fans who can easily buy without leaving Facebook, regardless of the intended purpose on Facebook, creating your online and mobile presence, contacting customers or encouraging people to take action has become a niche for the most diverse branches of commerce;

vii) **Pages:** these are intended for brands, businesses, organisations and public figures to create a presence on Facebook, whereas profiles represent individuals only. Anyone who has an account can create a page or help manage one if they are given a role on the page, such as administrator or editor;

viii) **Places:** tool for Facebook locations, it is possible to connect and manage all users' shops on Facebook, which allows to list all shops so that all people can locate them in search or when they arrive on their main business Page. This is especially important for people who use their mobile phones to find information when they are on the move;

ix) **Groups:** provide a space for people to chat about common interests. You can create groups for any purpose, such as family reunions, your sports team with co-workers, your book club, and customise the group's privacy settings according to who you want to join and see the group;

x) **Apps:** interact with Facebook's internal features, games like chess and *scrabble* among others are available to its users;

xi) **Events:** members inform their friends about upcoming events in their community, to organise social gatherings or simply to say what they are feeling at the moment. Events allow the user to organise and participate in real-world gatherings with people on Facebook. You can create or participate in an event of any kind, from a birthday dinner to a school fundraiser.

Figure 5 shows how these tools appear on the screen. The red highlight indicates the search options, which can be "everything" or "specific" (publications, people, photos, videos, shop, pages, places, groups, applications or events) from a keyword for example. We used the keyword "Astronomy" as an initial search option to find possible existing Facebook groups, which will be the data source for this work in the future.

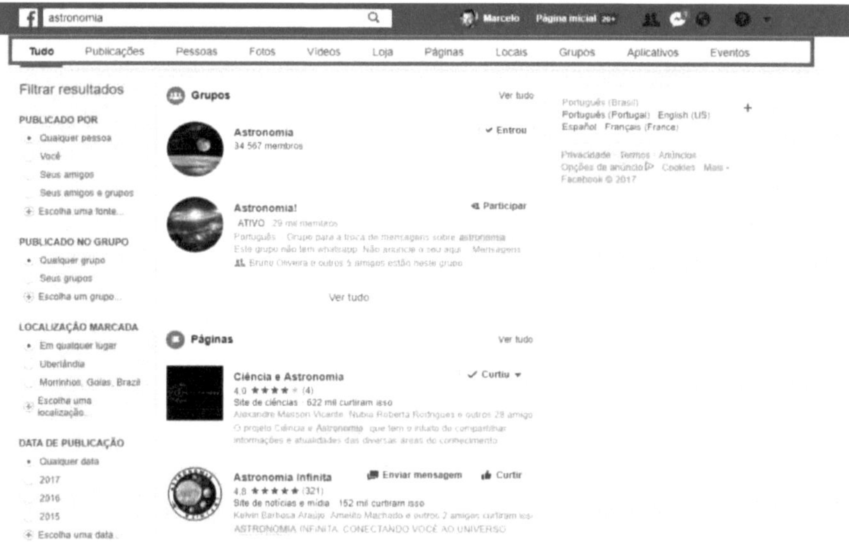


**Figure 5: Screen showing available search options.**
Source: Facebook (2017).

The search was conducted using as search terms: Universities, Centres, Museums, Colleges, Association and Astronomy. We identified the number of existing groups on Facebook/Brazil on 16 October 2017, the date of data collection. This is the data that made up our sample.

With this data saved, we created 6 categories, which served as support for analysis and understanding (SANTOS, ARANTES and USTRA, 2013). These categories correspond to the search terms. Data referring to federal, state and private universities will be called "UNIVERSITIES"; those linked to research centres and federal institutes will be called "CENTRES"; those referring to private and public colleges will be called "FACULTIES"; those belonging to permanent (building/place) and mobile (itinerant) museums will be called "MUSEUMS"; those composed of associations of public and private entities will be entitled "ASSOCIATIONS" and, finally, those composed of the most diverse groups, but with a common objective which is the dissemination, popularisation, sale and tips of astronomical equipment, news, events, forums, lectures, work shop and Astronomy courses will be entitled "MIXED".

When selecting the group tab, we see the following options, as highlighted in red in Figure 6.

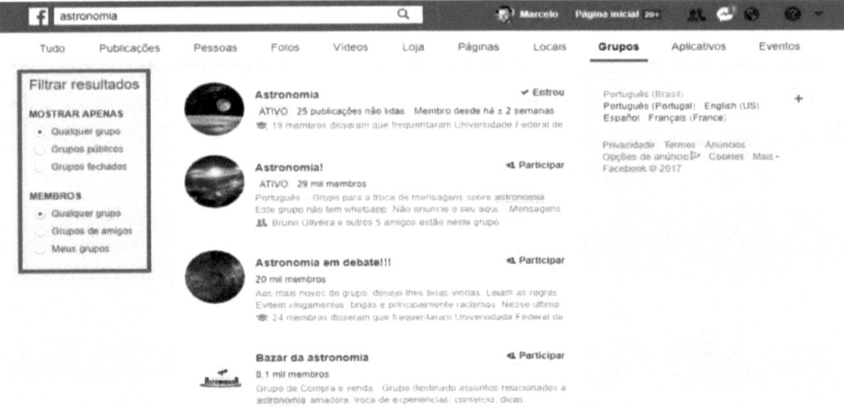


**Figure 6: Screen showing how to filter results**

Source: Facebook (2017).

We found two types of groups on Facebook: Closed Groups (GF) - anyone can find the group and see who is in it, but only members can post and see the group's posts, and Public Groups (GP) - anyone can post and see the group, its members and its posts.

To be able to identify the existing content in the FGs, we have to click on the participate option, as shown in Figure 7, and wait for the approval of the group administrator.

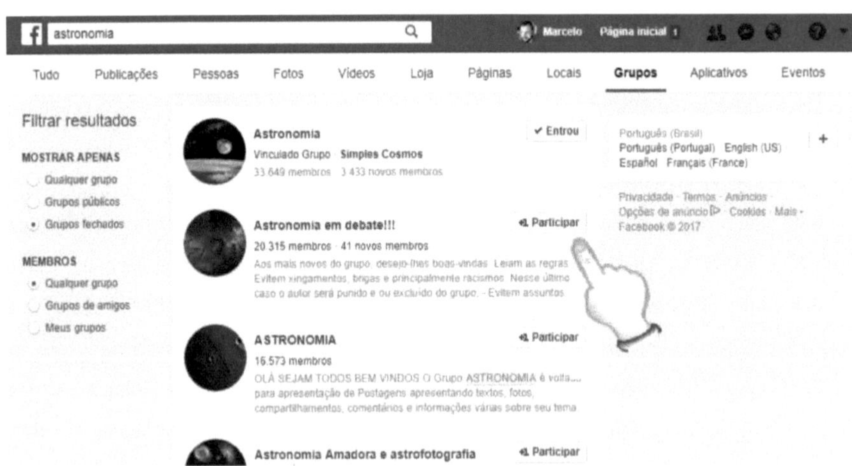

**Figure 7: Screen indicating the location for enrolment in Closed Groups.**
Source: Facebook (2017).

For GPs, when choosing the group, one can participate in discussions, check participating members, events, videos, photos and files among other options, as shown in Figure 8 highlighted in red.

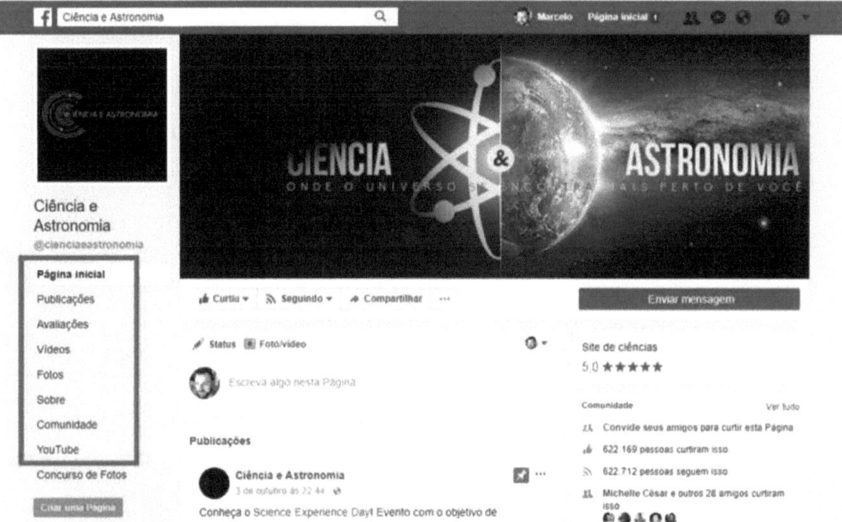


**Figure 8: Screen indicating the resource (interaction) options available in the group.**
Source: Facebook (2017).

All searches were carried out using the GP filter, in order not to have to wait for the approval of the group administrator, in this case GF, as the delay could jeopardise and delay data collection and the completion of this work.

We sought to identify the types and descriptions of each group within the categories listed (Universities, Centres, Colleges, Museums, Associations and Mixed). To do this, we proceeded as follows: we sought to identify which were the areas of activity and whether they are linked to any institution, NGO and/or GA highlighted in red in Figure 9.

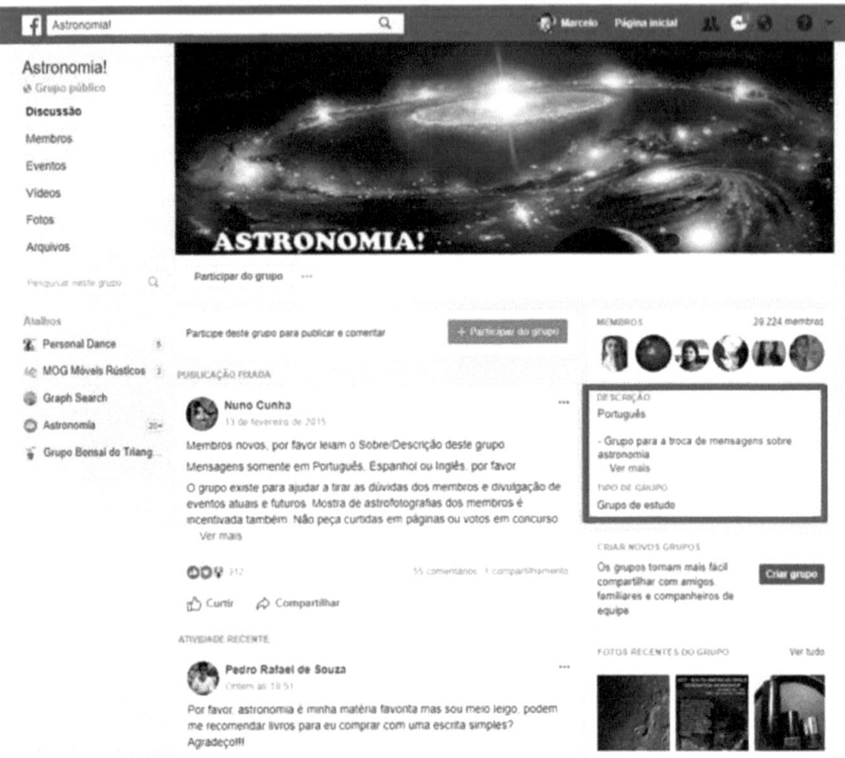


**Figure 9: Screen where there is a description of the group, area of activity and its purpose.**
Source: Facebook (2017).

After a successive search for Facebook groups related to Universities, Centres, Colleges, Museums, Associations and Mixed, we counted the number of members corresponding to each group, using the indications provided in front of the name of each group, as highlighted in red in Figure 10.

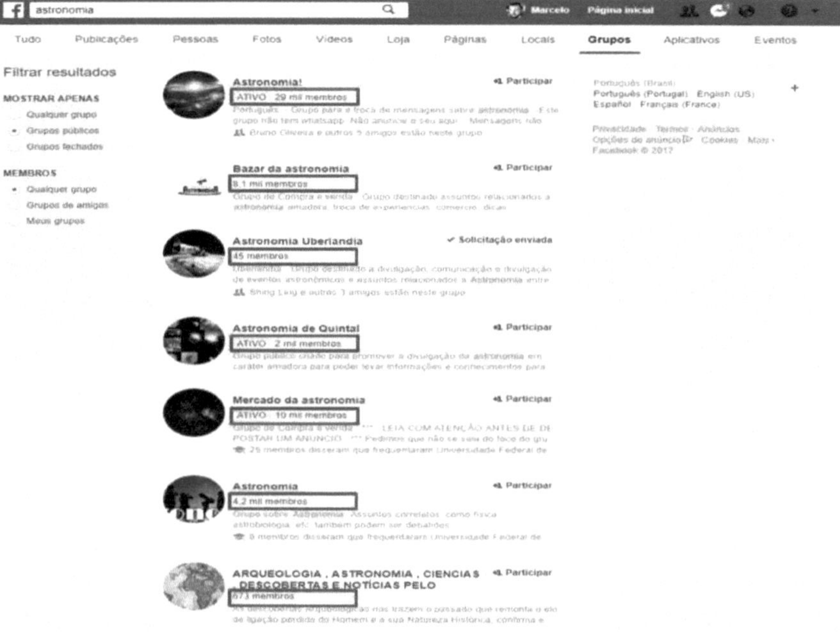


**Figure 10: Location showing the number of members in each GP.**

Source: Facebook (2017).

# CHAPTER 6

# DATA ANALYSIS

When quantifying the categories, we sought to identify the number of existing groups for each category on Facebook/Brazil as indicated in Figure 11, both belonging to FGs and GPs.

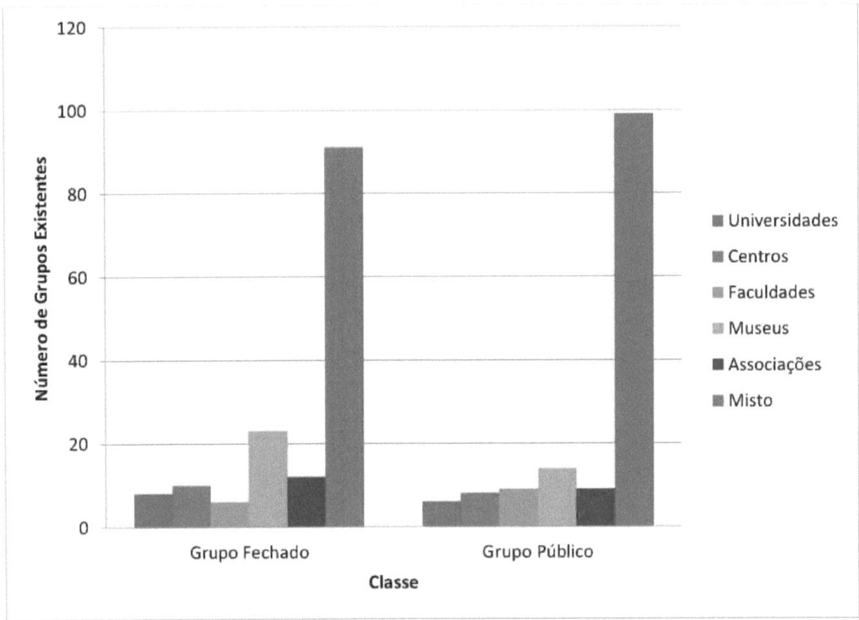


**Figure 11: Number of members in each category.**

Source: Facebook (2017).

Subsequently, we sought to identify the areas of activity of each category, as well as the institutions to which they are linked. The results are presented in Table 4.

**Table 4: Areas of activity of each category and the institutions that manage them.**

| Categories | Areas of Expertise | Linked to |
|---|---|---|
| **Universities** | Higher education in astronomy. | GA and NGOs |
| **Centres** | Publicising events, courses and curiosities. | NGOs |
| **Colleges** | Publicising events, sky observations, photos and news. | NGOs |
| **Museums** | Promotion of meetings, dissemination of photos from observations of the universe using the telescope. | GA and NGOs |
| **Associations** | Promotion of interaction, dissemination and planning of Astronomy activities. | NGOs |
| **Mixed** | Posts and publicises texts, photos, events, polls, activity planning, sharing, product sales, comments and topics involving: Cosmology, Astronautics, Astrophysics and Geography. | GA and NGOs |

Source: Facebook (2017).

Although we found some recurrence of members in other groups, that is, there were members who participated in more than one category and subcategory, considerably increasing the number of members to more than 6.5 million users connected to the analysed group, especially those contained in the Mixed category. For example, in a given group, there were more than 100,000 members participating in the group in question, we quantified the members participating in the categories according to Table 5, regardless of whether these members were present in other groups, categories or subcategories.

**Table 5: Number of participating members in each category.**

| | Categories | | | | | | |
|---|---|---|---|---|---|---|---|
| **Group** | **Universities** | **Centres** | **Colleges** | **Museums** | **Associations** | **Mixed** | **Total** |
| **Closed** | 49 | 48 | 27 | 1.419 | 873 | 225.677 | 228.093 |
| **Public** | 15 | 36 | 45 | 232 | 508 | 426.178 | 427.014 |

Source: Facebook (2017).

Facebook provides technological resources that can be widely used to popularise Astronomy, which are through: posting videos, events, polls, short and long courses, discussions, field activities, documents, symposia, meetings, photos and/or images, lectures, news, software, organisation of technical visits.

We classified and grouped these elements into the following subcategories, as schematised in Figure 12:

> **Dissemination** - activities, poster, course, meeting, event, lecture, symposium and visits;

> **Interaction** - discussion, dissemination, news, thoughts, quizzes and software;

> **Sharing** - documents, videos and photos;

> **Sales** - discussion, questionnaires and selling.

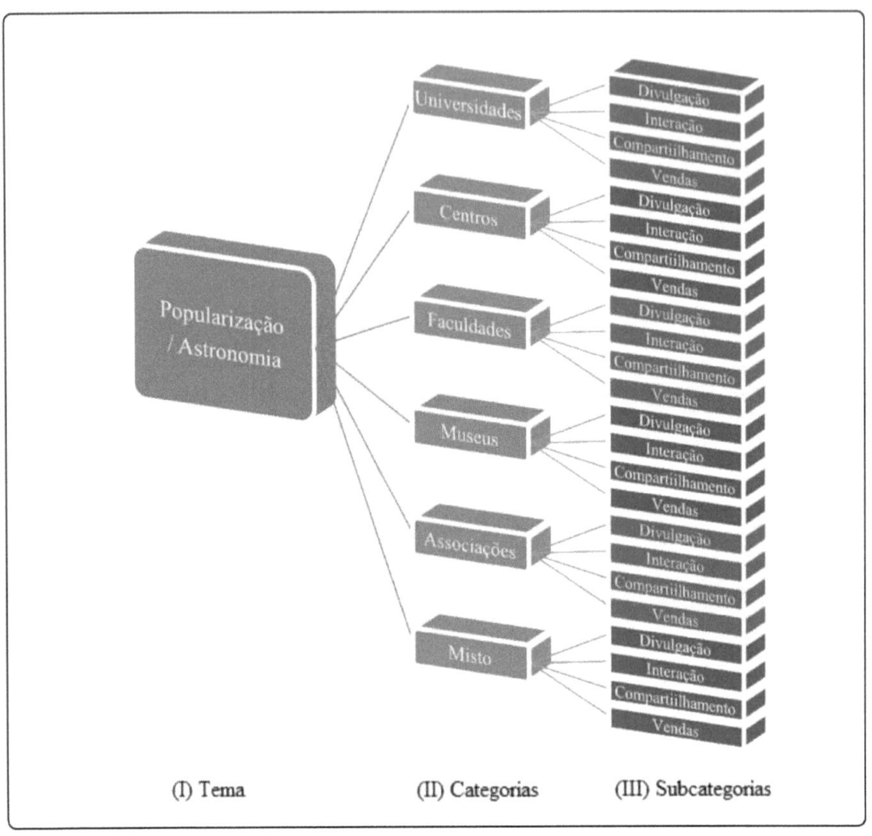


**Figure 12: Organisational chart of categories and subcategories.**

Source: Gomes (2017).

Regarding the popularisation of astronomy, we can say that it is broad, since in just over 4 months (first publication analysed is from 12 May 2017, in the Mixed category, the other categories are with later dates) we collected 1,246 data.

In the Dissemination subcategory, the activities developed in each group are diverse, comprising, for example, activities with schools, communities and asteroid observations, and the category that most promotes this type of activity is Centres, with 46%, in second place the category Colleges, with 21% and, in third, Museums, with 17%. The other categories represent less than 10 per cent of these activities involved, as indicated in Figure 13.

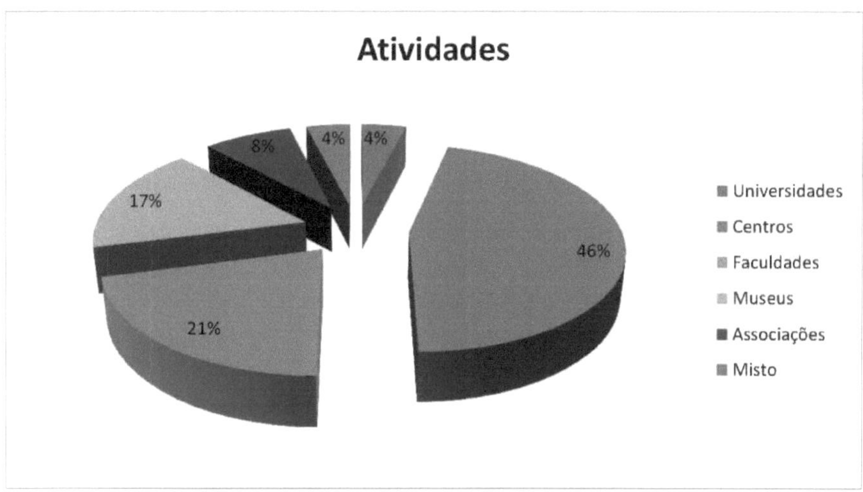


**Figure 13: Activities carried out in each category.**
Source: Gomes (2017).

Regarding the courses offered to its members, it is interesting to note that this category, in a way, is more distributed. The Associations category leads with 26%, Centres and Museums with 22%, Colleges 13%, Universities and Mixed with 9%. Both categories offer courses on astronomy, such as observations of the sky and on the theme "Solar System". We noticed that the Museums category did not advertise any symposium, while the Centres and Mixed categories did 27%, Universities and Associations 18% and Colleges 9%, according to Table 6.

**Table 6: Sharing/Dissemination of poster, course, meeting, lecture and symposium.**

| | Categories | | | | | |
|---|---|---|---|---|---|---|
| **Subcategory I** | **Universities** | **Centres** | **Colleges** | **Museums** | **Associations** | **Mixed** |
| **Poster** | 11% | 6% | 19% | 42% | 17% | 6% |
| **Course** | 9% | 22% | 13% | 22% | 26% | 9% |
| **Meeting** | 39% | 8% | 18% | 21% | 5% | 8% |
| **Lecture** | 10% | 3% | 7% | 27% | 10% | 33% |
| **Symposium** | 18% | 27% | 9% | 0% | 18% | 27% |

Source: Gomes (2017).

Regarding the promotion of meetings and events, as presented in Figure 14, Astronomy and Astrophotography present the highest percentage of publications, while the average of these events promoted by category, Misto presents 49%, Museums, Colleges and Associations together represent on average about 33.27% of meetings and events, according to Figure 15.

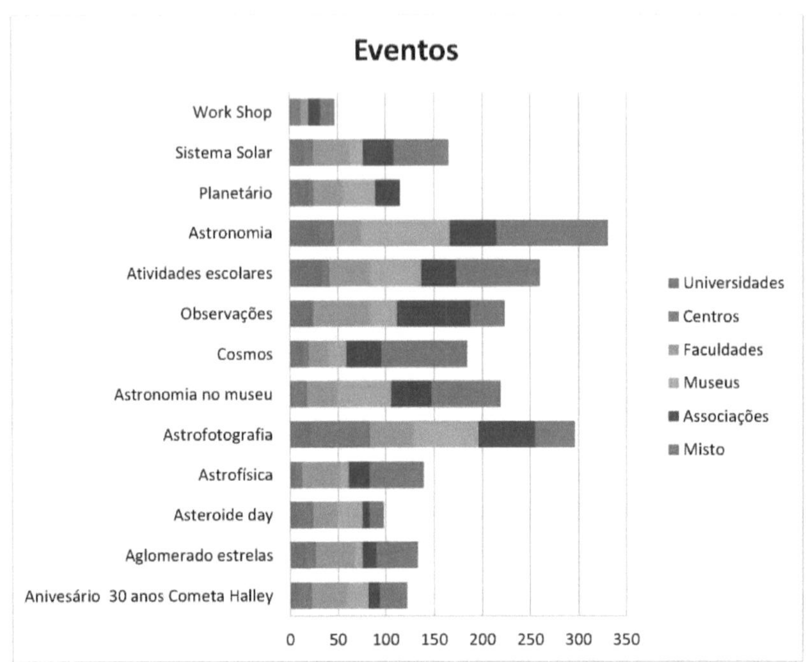


**Figure 14: Distribution of events in each category.**

Source: Gomes (2017).

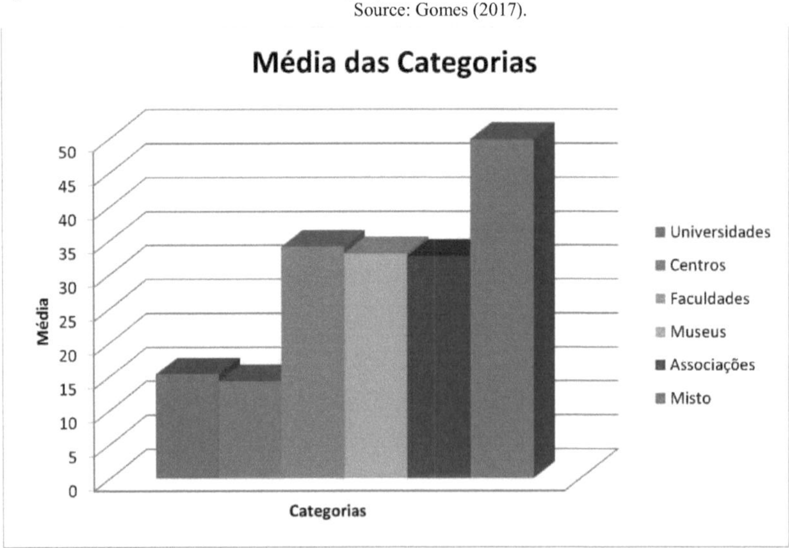


**Figure 15: Average of events and meetings promoted by the categories Universities, Centres, Colleges, Museums, Associations and Mixed.**

Source: Gomes (2017).

The subcategory Interaction is related to the dissemination of information in the sense that members can interact with the group (discussions), members participated more in the categories Dissemination and Centres, with 33 and 28% respectively. Most of this percentage was related to students from a state school in the state of São Paulo, focused on the Science Fair. The other categories highlighted general issues, such as the Chinese Station that would fall to Earth in 2018, Black Holes, among other subjects, as shown in Figures 16 and 17.

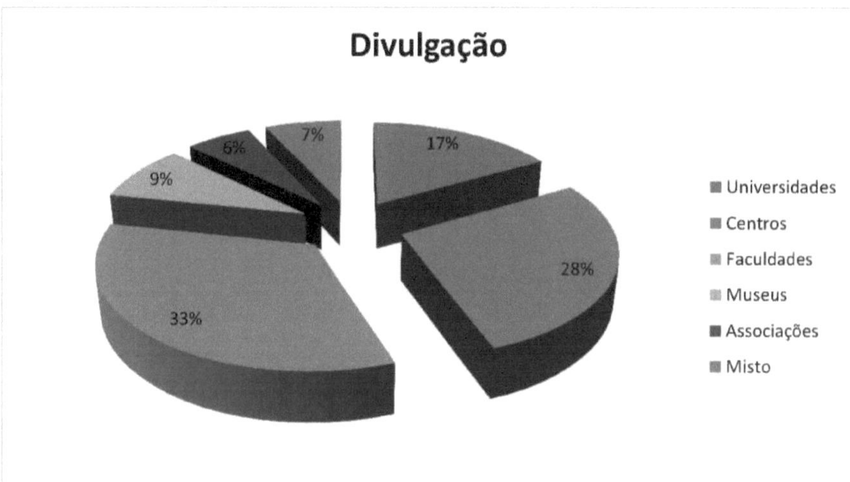


**Figure 16: Dissemination of curiosity and information on various events by group members.**

Source: Gomes (2017).

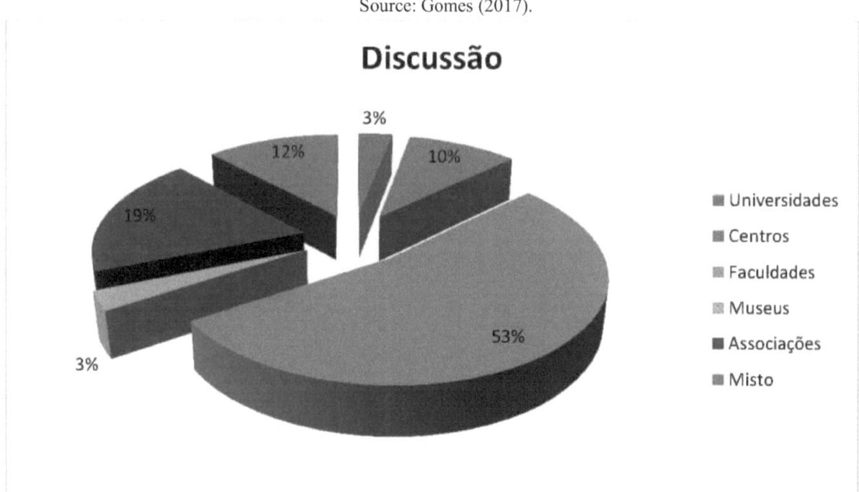


**Figure 17: Participatory discussion of members on some issue.**

Source: Gomes (2017).

The news disseminated in the groups mostly came from the Hubble Telescope, Cassini I and II Probe, Satellites and Terrestrial Telescopes, as well as from newspapers and magazines, and in the

Associations category this information corresponded to 31%, Museums 24%, Mixed 20%, Universities 12%, Centres 8% and Colleges 5%, as shown in Figure 18.

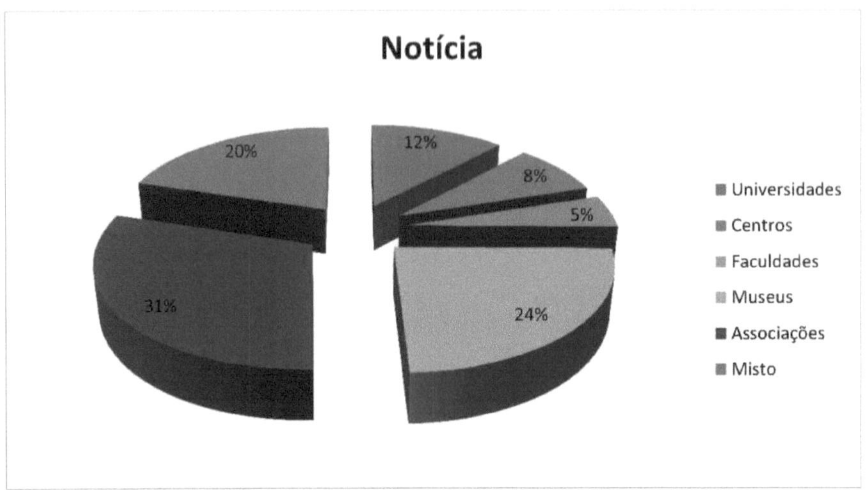


**Figure 18: News from newspapers, magazines and documentaries.**
Source: Gomes (2017).

To assist in observing the night sky, many groups used software, for example Stellarium, to locate and simulate which constellations would be visible on the intended day of observation, as well as locating galaxies and black holes. Another news item explored was the detection of gravitational waves, predicted by Albert Einstein, based on his theory of General Relativity, in 1916, and detected in 2015 by the Laser Interferometer Gravitational-Wave Observatory (LIGO) in Livingston, Louisiana, USA, and by the Virgo Interferometer in Cascina, Italy. The Mixed category makes no mention of software. The data are shown in Figure 19.

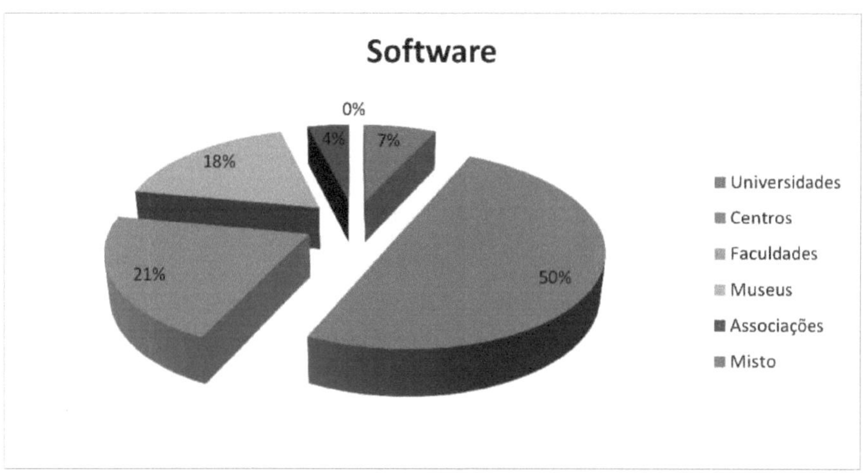


**Figure 19: Dissemination of software such as Stellarium.**
Source: Gomes (2017).

The subcategory Sharing is related to the sharing of the "printed" record. The Museums category had 40% of documents in PDF or DOC extension to be downloaded and indication of videos and *websites* to deepen or simply know a little more about what a black hole is, for example, in addition to bringing documentaries, as shown in Figure 20.

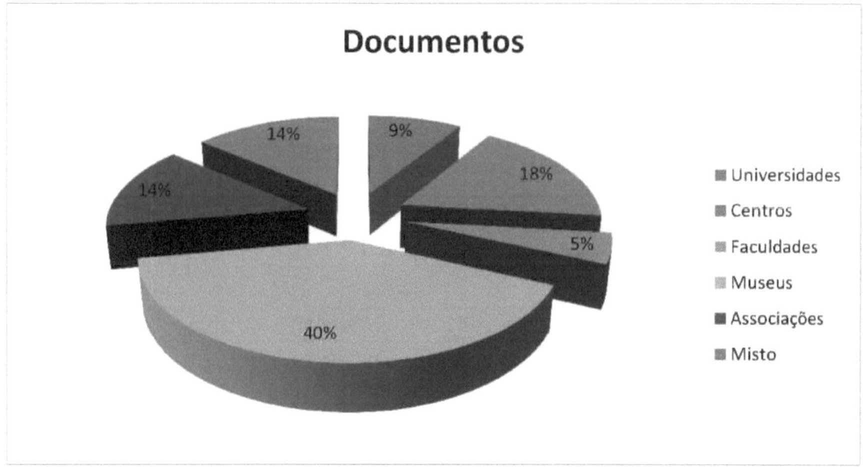


**Figure 20: Availability of documents as a research source.**
Source: Gomes (2017).

The posting/sharing of photos comes mostly from cameras, mobile phones, tablets, telescopes, with the Mixed category using 30% of these posts to popularise Astronomy and Associations 21%, Museums 18%, Centres 12%, Universities 10% and the Colleges category 9% respectively. The data are shown in Figure 21.

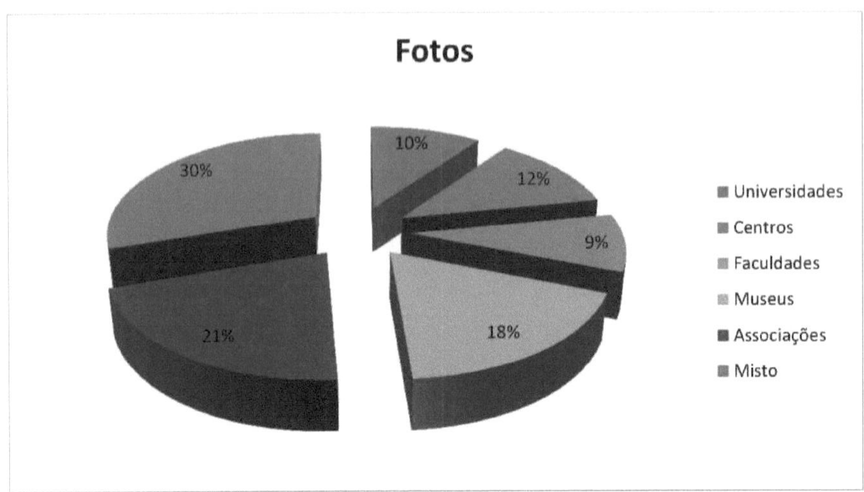


**Figure 21: Publication of photos from telescopes, cameras and satellites.**
Source: Gomes (2017).

Video sharing is an aspect that is more distributed across the categories, with the Centres category using 31% of these shares, Mixed representing 25% and the Associations, Museums, Colleges and Universities category representing 16%, 15% 7% and 6% respectively, as Figure 22 reveals.

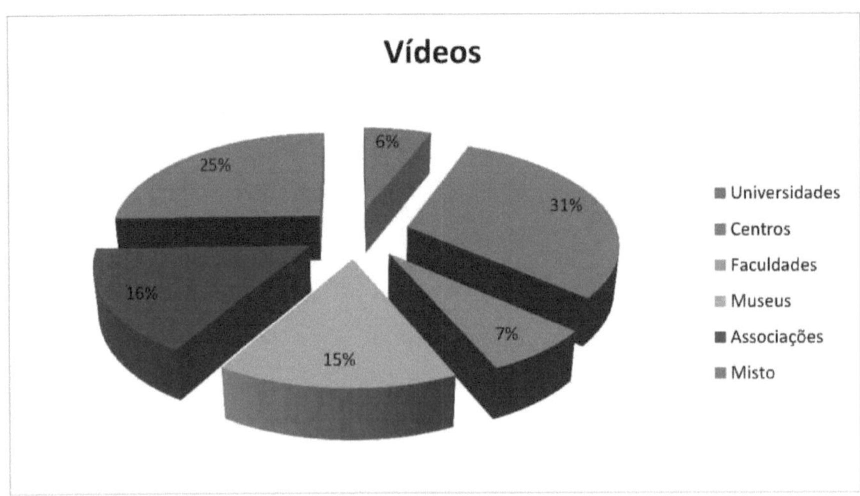


**Figure 22: Sharing documentaries, simulations and stellar events.**
Source: Gomes (2017).

Finally, subcategory IV sales is related to the sale of astronomical equipment, meteorites and indication of components for telescopes, such as type of lenses, cost benefit of telescopes, cameras, etc. In this subcategory, Misto leads with 33% of these sales and the category Colleges and Universities with 22%, mainly regarding the indication and/or discussion for the acquisition of certain equipment such as lenses, telescopes, lunettes, binoculars etc. The data are shown in Table 7.

**Table 7: Percentage of categories in relation to questionnaires, sales and discussions.**

| | Categories | | | | | |
|---|---|---|---|---|---|---|
| **Subcategory IV** | **Universities** | **Centres** | **Colleges** | **Museums** | **Associations** | **Mixed** |
| **Discussion** | 6% | 14% | 31% | 2% | 29% | 18% |
| **Questionnaires** | 9% | 5% | 18% | 41% | 5% | 23% |
| **Sales** | 22% | 11% | 22% | 0% | 11% | 33% |

Source: Gomes (2017).

As for the lowest frequency observed, the shares referring to Symposia, the Museums category did not make any citation to the subject, as well as Universities does not mention thoughts, the Mixed category does not make any publication of Software and the Museums category does not deal with sales in its publications. On the other hand, what is most published in the Mixed category is sales, with 33%, the Centres category is software, with 50%, Associations is thoughts, with 40%, and the Mixed category is thoughts with 27% according to Table 8.

**Table 8: Less frequent sharing.**

| | Categories | | | | | |
|---|---|---|---|---|---|---|
| **Sub-category IV** | **Universities** | **Centres** | **Colleges** | **Museums** | **Associations** | **Mixed** |
| **Symposium** | 18% | 27% | 9% | 0% | 18% | 27% |
| **Thoughts** | 0% | 10% | 30% | 0% | 40% | 20% |
| **Software** | 7% | 50% | 21% | 18% | 4% | 0% |
| **Sale** | 22% | 11% | 22% | 0% | 11% | 33% |

Source: Gomes (2017).

Facebook is accessed to carry out the most diverse activities, as shown in Figure 23. According to the data collected, the use of this technological application has been predominantly for sharing photos, news and disclosures.

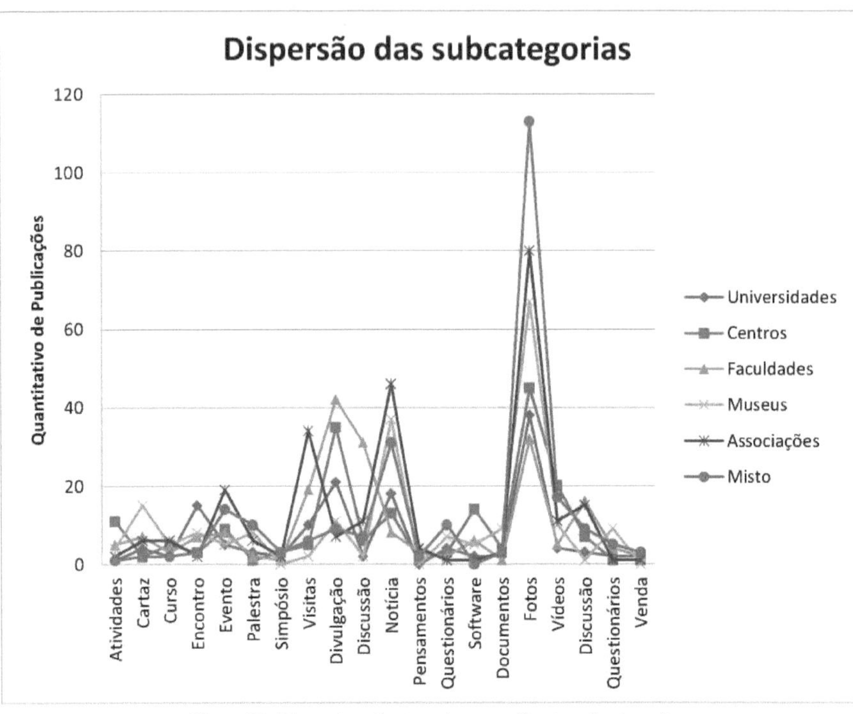


**Figure 23: Dispersion of categories according to subcategories.**

Source: Gomes (2017).

From these events presented in figure 23 above, we realise that a good part of these events are related to:

> Cosmology (a field that studies the origin, structure and evolution of the Universe through the application of scientific methods) which is sometimes confused with Astrophysics;

> Astrophysics (a field that studies the structure and properties of celestial objects and the universe as a whole through theoretical physics);

> Position Astronomy which the simple act of looking at the night sky, on a cloudless night and away from the city lights, can identify a large number of stars of varying brightness and colours, in addition to seeing a "shooting star", or suddenly witnessing some phenomenon in the sky (passage of a satellite, weather balloon or even an aeroplane, apparent change in brightness of a star, etc.), which in a way, has become a new area of study Astrophotography;

> Astro photography (specialised area of photography that involves, recording/photographing images of celestial bodies over large areas of the night sky. As well as being able to record the details of large bodies such as the Moon, Sun and planets). Astrophotography has the ability to show objects invisible to the human eye, such as Nebulae and Galaxies. This is done

by long exposure, using an ability that both films and digital sensors have to accumulate photons of light over long periods of time;

> Extragalaxies (area that studies other galaxies, planets and stars that are outside the Milky Way);

> Radio astronomy (a field that studies the electromagnetic radiation emitted or reflected by celestial bodies, the reception of this electromagnetic radiation, this detection is done by means of radio telescopes).

# CHAPTER 7

# CONCLUSIONS

The main purpose of all the analysis carried out in this work was to study the potential of Facebook as a tool for popularising astronomy. We identified 150 Closed Groups, of which we did not obtain any data because they are private, and 145 Public Groups, composed of 421,017 active members. When we add Closed Groups and Public Groups we have a total of 295 existing groups on Facebook/Brazil that have Astronomy as their central theme, and more than 655,100 members who are directly or indirectly interested in Astronomy. They reveal that they seek to popularise it, so that more people can have access to the various information in this field of knowledge. This information can range from photos of the Moon, for example, to the detection of gravitational waves. The number of Closed and Public Groups needs to be studied, since new groups are created every day for the most varied purposes, such as a specific event.

During the research we realised that the Astronomy Groups are expressive, if analysed in relation to the number of members, which are more than 655.1 thousand members.

We found that the Popularisation of Astronomy through social networks operates on several fronts, in which some groups have focused their action on the following sub-areas: Astrophysics, Position Astronomy, Extragalactic, Radio Astronomy, Cosmology.

According to our research, the sub-area that stood out the most was Cosmology, followed by Astronomy and Extragalactics, but we noticed a trend in Astrophotography, as many are increasingly looking for sophisticated equipment in order to obtain "perfect" images or even images from the Hubble Telescope and the Cassine and Voyager probes. There are several discussions about the best type of image treatment, filters, in short, a significant part of the groups / people have this purpose. It should be emphasised that, in general, both sub-areas were present in all groups, even if in small percentages.

As a result of the participation of the members in the groups, we found that about 37% of the members, being the ones who keep the group running with constant publications to the other members of the group, informing them of what is happening about Astronomy, which can be: posters, courses, events, lectures, etc. The trend is that this number may be even higher, since users with active Facebook accounts represent about 66.92% of the Brazilian population, thus having a huge audience waiting for knowledge, whatever it may be.

Some groups have greater prominence in sales, others in posting videos and photos, but in general, the groups allow their members a space conducive to the discussion of the most varied subjects. A large part of the groups analysed are linked to Government Agencies, being them public, that is, Public Universities, Museums, Institutes, Centres, Schools, where teachers/researchers/disseminators of Astronomy have endeavoured and at the same time ventured into the creation/participation of groups with the purpose of popularising Astronomy, sharing the knowledge acquired over the years. The remainder is composed of non-profit entities, NGOs, which have partnerships with Museums, Centres, Colleges and Universities and, finally, in a minority, Journalists, trained in scientific dissemination and who come, for the most part, from the José Reis Nucleus of Scientific Journalism (NJR), of the University of São Paulo (USP) and the Open Laboratory of Interactivity for the Dissemination of Scientific and Technological Knowledge (LAbI), of the Federal University of São Carlos (UFSCAR). They understood the relevance of the need for training and specialisation of journalists in scientific dissemination/popularisation.

The resources that predominate as a tool for Popularising Astronomy on Facebook are the sharing of photos, news, disclosures, visits, videos and events, which together account for 67% of the publications made in the groups. However, the publication that wins in percentage are the photos, which become accessible, as they do not require sophisticated equipment. However, there are those who seek to make considerations about the images obtained and posted, and seek information such as, for example, image processing with appropriate software. This reveals a demand for astrophotography courses.

The tools offered by Facebook, a priori, meet the needs of the Popularisation of Astronomy, since the network is complex, contemplating the most varied audiences, but in general it causes a certain admiration for something that we do not see with the naked eye, but that through adequate equipment we obtain true works of art, such as the Rosette Nebula.

Despite being restricted to Facebook users, this tool can and is a complementary educational tool, as it allows knowledge to be aggregated by people who sympathise with Astronomy, thus making knowledge more enjoyable, that is, not forced. In addition, Facebook provides tools such as Wall, Events, Applications, among others, which makes it possible for people who do not attend school environments to obtain this information, which helps in science education, expanding and, at the same time, making the subject popular, not being treated as something surreal or done only and exclusively in highly complex laboratories.

## References

ABCMC. 2006. **Brazilian Association of Science Centres and Museums**. Available at <http://www.abcmc.org.br/publique1/cgi/cgilua.exe/sys/start.htm?infoid=39&sid=18>, accessed in Aug. 2016.

ALBAGLI, S. **Divulgação Científica: Scientific Information for Citizenship**. Ciência da

Informação, v. 25, n. 3, 1996.

AQUINO, A.; BRITO, A. **Study of the Feasibility of Using Facebook for Education**. In: WORKSHOP ON INFORMATICS EDUCATION, UFPR 2012.

BARDIN, L. **Content Analysis**. São Paulo: Editions 70, 2011.

BRASIL. **Art. 1° The National Science and Technology Week is hereby instituted, to be celebrated in the month of October of each year**. Brasília, DF, 9 June 2004.

BRETONES, P. S. **The Secrets of the Solar System**. São Paulo - SP: Atual, 1993.

BUENO, C.; DIAS, S. **The Act of Publicising as a Training Laboratory**. Com Ciência, n. 100, p. 0-0, 2008.

CALIPO, V. **Youth and the Internet Age: Integration and Interaction**. UMESP. XXXI Intercom, 2008.

CASTELLS, M. **The Internet Galaxy**. [S.l.]: Rio de Janeiro. Jorge Zahar editora, 2004.

CREATIVE, C. **Mark Zuckerberg**. 2016. Available at: <https://pt.wikipedia.org/wiki/Mark_Zuckerberg>, accessed Aug 2017.

CREATIVE, C. **Social Networking**. 2017. Available at: <https://pt.wikipedia.org/wiki/Rede_social>, accessed Sep 2017.

CRUZ, A. F. et al. **Social Network: Potentialities of Facebook for Face-to-Face Education of the Degree in Pedagogy**. EDUCA-Multidisciplinary Journal in Education, v. 1, n. 1, p. 39-55, 2014.

EBC, **Social Mobilisation of June 2013 Led the Government to Propose Five Pacts**. Available at < http://agenciabrasil.ebc.com.br/geral/noticia/2014-06/mobilizacao-social-de- june-2013-led-government-to-propose-five-pacts>, accessed Jun 2017.

EBC. **Facebook Reaches 1 Billion Users Worldwide**. Portal EBC, Available at: <http://www.ebc.com.br/tecnologia/2012/10/facebook-chega-a-1-bilhao-de-usuarios-no- world>, accessed in May 2017.

FACEBOOK. **Notes Mark Zuckerberg**. Available at <https://www.facebook.com/notes/mark-zuckerberg/bringing-the-world-closer-together/10154944663901634/>, accessed May 2017.

FACEBOOK. Available at <http://www.facebook.com>, accessed Aug 2015.

FERREIRA, J. L.; CORRÊA, B. R. P. G.; TORRES, P. L. **The Pedagogical Use of the Social Network Facebook**. Colabor@-A Digital Journal of CVA-RICESU, v. 7, n. 28, 2013.

FREIRE, P. **Extension or Communication?** Translation: Rosisca Darcy de Oliveira, 10. ed. Rio de Janeiro: Paz e Terra, 1992.

GERMANO; M. G.; KULESZA, W. A. **Popularisation of Science: A Conceptual Review**. Caderno Brasileiro de Ensino de Física, v. 24, n. 1, p. 7-25, 2008.

GERMANO, M. G. **A New Science for a New Common Sense**. EDUEPB, 2011.

GIDDENS, A. **The World in the Age of Globalisation**. Editorial Presença, 2013.

GOMES, M. O. **Social Networks in Physics Teaching: Facebook**. II Meeting Interinstitutional of PIBID - Uberlândia, MG. 2013.

GONÇALVES, P. **Educational Utilisation of Facebook in Higher Education**. University of Évora, p. 3, 2010.

GOHN, M. da G. **Sociology of Social Movements**. Cortez Editora, 2016.

LÉVY, P. **Collective Intelligence (A)**. [S.l.]: Editions Loyola, 2007.

LOPES, M. M. **Brazil Discovers Scientific Research:** Museums and the Natural Sciences in the 19th Century. São Paulo: Editora Hucitec, 369p. 1997.

MARANDINO, M. et al. **Non-formal education and scientific dissemination: what those who do it think**. Proceedings of the IV National Meeting on Research in Science Teaching, 2004.

MARTINEZ, E. **La Piramide de la Popularización de La Ciencia y La Tecnología**. In: MARTINEZ, E.; FLOREZ, J. (comp.) **La Popularización de La Ciencia y La Tecnología**. Mexico: FCE-Unesco-Red-POP FCE, p. 9-16, 1997.

MARTINS, I. A. **Informatics in education:** a path of possibilities and challenges. Serra: IFES Publisher, 2010.

MELO, A. C.; DIEGUES F. **On the Launch Pad**. ABCMC in the media. 2002.

MONTEIRO, S. D. **The Cyberspace: The Term, The Definition and The Concept**. Datagramazero, Rio de Janeiro, v. 8, n. 3, p. 0-1001, 2009.

MOTTA, G. R.; GAVA, T. R. **Virtual Learning Communities as a Teacher Training Space**. NOBRE, IA [et al.] Informática na educação: um caminho de

possibilities and challenges. Federal Institute of Education, Science and Technology of Espírito Santo. Serra-ES, 2011.

MUELLER, M. S. **Popularisation of Scientific Knowledge**. Journal of Science and Information, v. 3 n. 2, Apr. 2002. Available at: <http://www.dgz.org.br/abr02/Art_03.htm>, accessed in July 2017.

NASCIMENTO, S. S. **The Image in the Popularisation of Science:** Astronomy. Thematic Area: Visual Education, Visual Languages and Art. 2008.

NAVAS, A. M. **Conceptions of Popularisation of Science and Technology in Political Discourse: Impacts on Science Museums**. Doctoral Thesis, Faculty of Education-USP, 2008.

PAPERT, S. **The Children's Machine: Rethinking School in The Age of The Computer**. [S.l.]: Basic Books, 1993.

PERON, A. **Facebook Marketing - 2016 Data from the World's Largest Social Network**. Available at: <http://www.allanperon.com.br/facebook-marketing/#ixzz4tq5Nnoix>, accessed June 2017.

REIS, J. Point of view: José Reis. In L. Massarani, I. C. Moreira & F. Brito (Eds.), **Ciência e Público: Caminhos da Divulgação Científica no Brasil (pp. 73-78)**. Rio de Janeiro, RJ: Cultural Centre for Science and Technology of the Federal University of Rio de Janeiro. 2002.

SANTOS, N. S.; ARANTES, E. A. S.; USTRA, S. R. V. **What Motivates Students in a Physics Class?** São Paulo, SP. XX National Symposium on Physics Teaching - SNEF 2013.

SILVA, A. H.; FOSSÁ, M. I. T. **Content Analysis: Example of Application of the Technique for Qualitative Data Analysis**. Qualitas Electronic Journal, v. 16, n. 1, 2015.

SILVA, M. R. da; CARNEIRO, M. H. da S. **Popularisation of Science**: Analysis of a Non-Formal Teaching Situation. Annual Meeting of the National Association of Graduate Studies and Research in Education, p. 1-16, 2006.

SMITH, M. **Non-formal Education**. Infed. 2001. Available at: <www.infed.org/index.htm>, accessed Sep. 2017.

SOCIALBACKERS. **10 Fastest Growing Countries on Facebook in 2012**. Available at: <https://www.socialbakers.com/blog/1290-10-fastest-growing-countries-on-facebook-in- 2012>, accessed Aug. 2017.

STATISTA. **Countries Based on Number of Facebook Users**. Available at: <https://www.statista.com/statistics/268136/top-15-countries-based-on-number-of-facebook-users/>, accessed Jul 2017.

TOMAÉL, A. D. C. **From Social Networks to Innovation**. Ciência da Informação, Brasília, SciELO Brasil, v. 34, n. 2, p. 93-104, 2005.

WORLDOMETERS. **World Population 2017**. Available at: <http://www.worldometers.info/br>, accessed Sep 2017.

I want morebooks!

Buy your books fast and straightforward online - at one of world's fastest growing online book stores! Environmentally sound due to Print-on-Demand technologies.

Buy your books online at
**www.morebooks.shop**

Kaufen Sie Ihre Bücher schnell und unkompliziert online – auf einer der am schnellsten wachsenden Buchhandelsplattformen weltweit! Dank Print-On-Demand umwelt- und ressourcenschonend produzi ert.

Bücher schneller online kaufen
**www.morebooks.shop**

Printed by Books on Demand GmbH, Norderstedt / Germany